Exploring Science

The Exploring Science series is designed to meet all the Attainment Targets in the National Science Curriculum for levels 3 to 6. The topics in each book are divided into knowledge and understanding sections, followed by exploration. Carefully planned Test Yourself questions at the end of each topic ensure that the student has mastered the appropriate level of attainment specified in the Curriculum.

Exploring Science

Electricity
Energy Sources
Light
Magnets
Ourselves
Plants
Soil and Rocks
Sound
Uses of Energy
Weather

Cover illustrations:
Left An artist's impression of the human skeleton inside the body of a walking girl.
Above right A doctor listens to a baby's heartbeat through a stethoscope.
Below right A diagram to show the human heart.

Frontispiece If you fall and cut yourself, white blood cells rush to the wound to fight any bacteria entering the bloodstream through the broken skin.

Editor: Elizabeth Spiers
Series designer: Ross George

First published in 1990 by
Wayland (Publishers) Ltd
61 Western Road, Hove
East Sussex BN3 1JD, England

© Copyright 1990 Wayland (Publishers) Ltd

British Library Cataloguing in Publication Data
Catherall, Ed *1931–*
   Exploring ourselves.
   1. Man. Body
   I. Title II. Series
   612

ISBN 1-85210-912-2

Typeset by Nicola Taylor, Wayland
Printed in Italy by G. Canale & C.S.p.A., Turin
Bound in France by A.G.M.

Contents

Your body 6
Cells 8
Your skeleton 10
Bones and joints 12
Muscles and movement 14
Your diet 16
Your teeth 20
Your digestive system 22
Food absorption 24
Your blood 26
Your circulatory system 28
Lungs and breathing 30
Your brain 32
Your nervous system 34
Your skin 36
Your senses 38
Hormones 40
Reproduction 42
Leading a healthy life 44

Glossary 46

Books to read 47

Index 48

EXPLORING OURSELVES

Ed Catherall

YOUR BODY

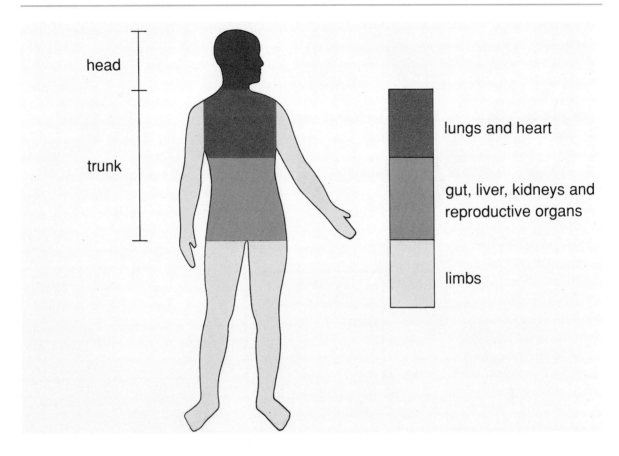

Your body is the most important thing in your life. It is very complicated, and scientists spend a lot of time and money finding out how it works. There are many tasks that it has to perform: it must eat, breathe, grow, be able to repair itself, move and get rid of waste and harmful substances. It must protect itself from illness and damage. It must be able to find out what is happening around it. It must be able to think and learn, and to reproduce itself. Because of this, it must be looked after very carefully.

Your body consists of a head, trunk and limbs. Your head contains some extremely important parts. One of these parts, your brain, controls your entire body. The air that you breathe and the food that you eat are taken in through your head. You communicate mostly with your head: by speaking, by the expressions on your face and by hearing. You also find out about your environment with your head, by smell, sight and taste.

Your trunk is divided into two sections. The upper half contains the lungs and heart, and the lower part contains the gut, liver, kidneys and reproductive organs. Your four limbs are for movement and receiving information by touch. They also help you to balance.

ACTIVITY

YOUR BODY SHAPE

YOU NEED

- **a large sheet of paper**
- **wax crayons**
- **a tape measure**

1. Lie on your back on the sheet of paper.
2. Ask a friend to draw around your body.
3. Use the tape measure to find out the length and width of your head and trunk from the outline. Record these measurements.

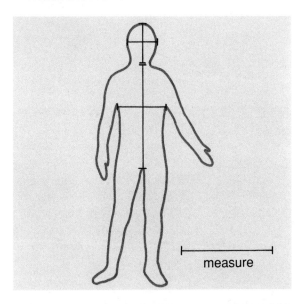

4. Measure the length and width of your arms and legs from the outline.

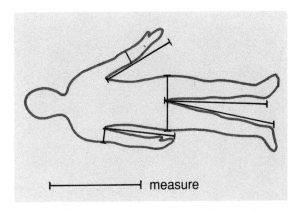

5. Compare these measurements with those of several friends. Make a set of bar graphs to show how the measurements vary. You could have a different bar graph for each type of measurement.

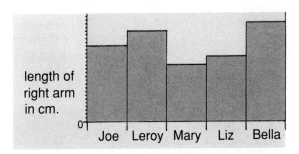

6. Study your bar graphs. Look for patterns in your results: for example, does the person with the largest head have the longest arms?

TEST YOURSELF

1. Which parts of your head help you to find out about your environment?
2. Name six tasks that your body must be able to perform.
3. How many pairs of limbs do you have? Name something that your upper limbs can do that your lower limbs cannot.

CELLS

Your body is made up of tiny 'building blocks' called cells. An adult has more than one trillion of these tiny cells. There are many different kinds of cell, each with its own special job to do, but they are all able to perform the same basic tasks. They are able to take in food to produce energy for your body to work, to make materials called proteins for repair and growth, and to reproduce (make copies of themselves). The different types of cell are arranged to make tissues and organs. For example, some cells form the pipes that make up your blood vessels, while others form your muscles.

Most of your cells do not last for the whole of your life. They die and have to be replaced. The only ones that cannot reproduce and replace themselves are your nerve cells (see page 35). They die off gradually, so that by the time you die, you have far fewer nerve cells than when you were born. Some cells, such as the ones on the surface of your skin, live for only a few days. Your red blood cells (see page 26) last for about four months, while bone cells may live for twenty-five years.

Children have far fewer cells than adults. As you grow, more and more cells are added to those you already have. Each of your cells, except for your sex cells (see page 42) is able to make an exact copy of itself, by dividing into two. When you are fully grown, your cells reproduce themselves mainly to replace those that die.

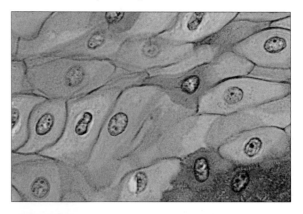

Above Cells from the tissue that covers the outer layers of the body. This type of tissue also lines the tubes inside your body, such as the gut.

Left A cell found in part of the brain.

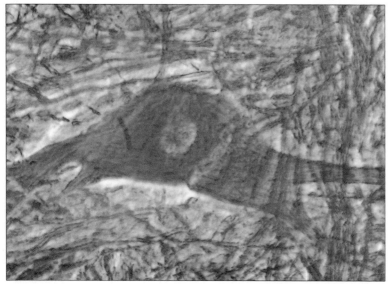

ACTIVITY

LOOKING AT CELLS

> YOU NEED
>
> - **a microscope**
> - **2 microscope slides**
> - **glass cover slips**
> - **a dropper (pipette)**
> - **an onion**
> - **a clean teaspoon**
> - **a sharp knife**

> WARNING: you must get an adult to help you when using the knife. You must also take great care with the glass slides and cover slips.

1. Rinse out your mouth with water.
2. Scrape the edge of the spoon gently over the inside of your cheek.
3. Wipe the edge of the spoon on your slide, to transfer what you have scraped off.
4. Cover the scrapings with a cover slip.

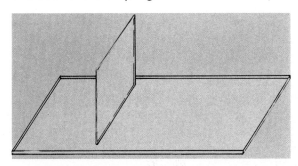

5. Look at your mouth scrapings under the microscope. If you use the strongest lens, you should be able to see some small, roundish objects. These are cells from the inside of your cheek. Draw what you see.

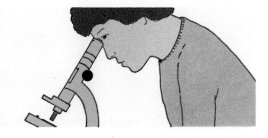

6. Ask an adult to cut the onion and peel off a very thin layer. Place a little of this on your other slide.
7. Use the pipette to put a drop of water on to the onion.

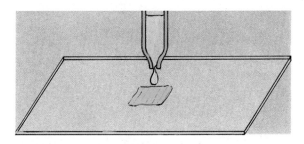

8. Cover the onion and water with a cover slip.
9. Look at your onion through the microscope. You should see something that looks like bricks in a wall. These are the cells that make up your onion. Draw them.

> ## TEST YOURSELF
>
> 1. What are most cells able to do?
> 2. How long do red blood cells live?
> 3. Which cells cannot replace themselves?

YOUR SKELETON

Inside your body, you have a skeleton made of bones joined together. There are three main reasons why you have a skeleton. Firstly, you are quite a large animal, so you need a strong frame to support your body. Secondly, your body contains many soft, delicate parts called organs and tissues. They need protection from injury. The skeleton is also used to hold the muscles in place, so that your body can move (see page 14). The skeleton also performs two other functions. It provides a place where blood cells can be made, and it acts as a store for a substance your body needs, called calcium (see page 18).

Your skull is made from several bones, all joined together to make a box to protect your brain and eyeballs. If you have a skeleton at school, look at the skull and you will see these pieces of bone, fitted together rather like a jigsaw puzzle. In your trunk, you will be able to feel your ribcage, pelvis (hips), shoulders and spine (backbone). The ribcage protects your heart and lungs, while your pelvis helps to support the weight of your body. It also protects some of the soft organs low down in your trunk. Your spine is the main support for your body. It is made up of a column of bones called vertebrae, and this helps to make it flexible (bendy). Imagine trying to do up your shoelaces if your spine was made of just one long bone! Your spine also protects your spinal cord, which links the brain with the rest of your nervous system (see page 32).

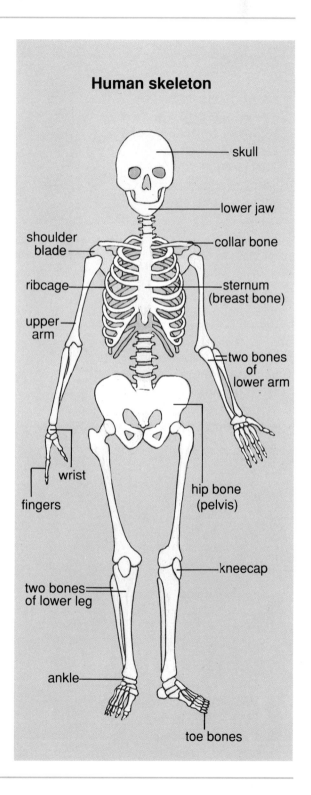

A diagram of the human skeleton.

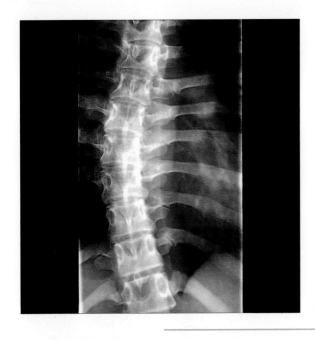

An X-ray of part of the spine, showing the vertebrae and part of the spinal cord.

Your arms are called upper limbs, and are attached to the spine by the shoulder blades. At the ends of your arms, your hands are attached by the wrists. There is a large number of bones in your wrists and hands; like the spine, this helps to make them flexible, so that they can perform complicated movements, such as writing. Your legs (lower limbs) are also connected to the spine, by your pelvis. Your feet are attached to your legs by the ankles. Again, the ankles and feet have many small bones, so that they are flexible.

ACTIVITY

YOU NEED

- **the body outline that you used on page 7**
- **a diagram of the human skeleton**

1 Work with a friend.
2 Use the diagram of the human skeleton to draw the positions of your bones inside your outline.
3 Label the bones. Choose different colours for bones that protect, limbs, and bones that support.

Remember that some bones protect and support. Use another colour to label these.
4 Make a key (guide) to show what the different colours mean.

Key	
■	Limbs
■	Limbs that support
■	Bones that protect
■	Bones that protect and support

TEST YOURSELF

1. Give three reasons why you need a skeleton.
2. Which bones protect your heart and lungs?
3. What are vertebrae?
4. Why is the wrist such a flexible joint?

BONES AND JOINTS

You have three main types of bone: long, short, and flat. They are all covered by a hard shell and are soft and spongy inside. Long bones are found in your arms, legs, fingers and toes. Inside, they have a substance called marrow, which makes red and some white blood cells (see page 26). Your backbone, ankles and wrists have short bones. These contain a network of long, thin fibres, rather like girders in a building. Flat bones make up your skull, ribs and shoulder blades, and have a thin spongy layer in the middle.

The bones in your skeleton have to be joined together and to move against each other without being damaged. The joints are specially designed for these tasks. There are several different kinds of joint. Most of them are either hinge joints or ball-and-socket. Hinge joints are found in places like the elbow and fingers. They allow movement mostly in one direction. Ball-and-socket joints are found in the hips and shoulders. They allow much more movement. There are also other kinds, such as the joints found between the bones in your spine. These bones have discs of spongy tissue between them that are like shock absorbers in a car. Another type allows almost no movement, and they are found between the bones of your skull and between the two bones in your forearm.

Joints that allow a lot of movement need to be protected from wear and tear. The ends of the bones are covered in a smooth layer called

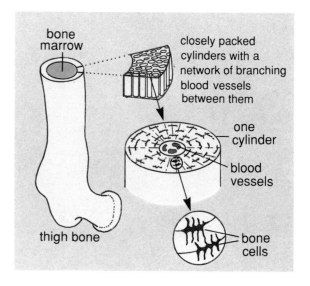

A cross-section through the thigh bone.

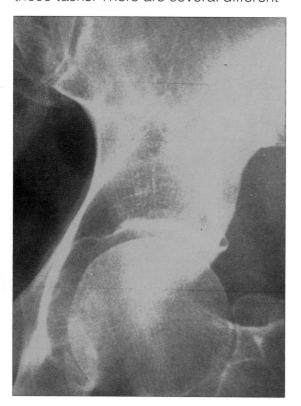

An X-ray taken through the pelvis. It shows the ball-and-socket joint, where the top of the thigh bone fits in.

cartilage. The whole joint is covered in an envelope, or capsule, which contains thick fibres called ligaments. They help to strengthen the capsule, to stop the bones from moving out of position. The capsule secretes (gives out) a slippery fluid that lubricates the joint, rather like oil on a bicycle chain.

ACTIVITY

INVESTIGATING YOUR JOINTS

WARNING: be careful when making the movements in this activity. Joints are easily damaged.

1 Hold one of your arms straight, away from your body. Turn the palm of your hand up.
2 Bend your elbow, then straighten it. How far does it move? Can you move your forearm from side to side?

3 Sit on a chair. Stretch your leg out straight in front of you. How far can you bend your knee? Does it allow you to move your lower leg from side to side?

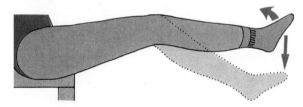

4 Stand up. How far can you swing your leg backwards and forwards? Can you swing it from side to side?
5 Test your shoulder joint by swinging your arm round. What sorts of movements can you make?

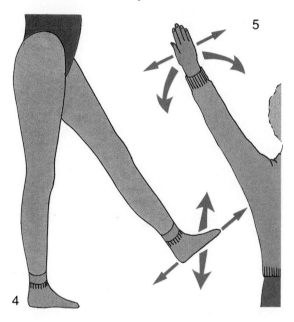

6 What sorts of movement can you make with your fingers?
7 Make a chart, showing which joint you have tested, what sort of joint it is and how you can move it.

TEST YOURSELF

1. What are the three main types of bone?
2. Where would you find joints that allow almost no movement?
3. How are joints protected from damage?

MUSCLES AND MOVEMENT

You need to be able to make a large range of movements in your everyday life. Muscles help you to do this. You have two main types in your body: involuntary and voluntary. Involuntary muscles are sometimes called automatic muscles. This is because you have no control over them. Perhaps you do not even know that they are there. Your heart (see page 29) has involuntary muscles to keep it beating throughout your lifetime. Imagine what life would be like if you had to remember to keep your heart beating! Your food is helped down your digestive system (see page 22) by involuntary muscles.

Voluntary muscles are those that your body deliberately controls, even though you may not think about it. They are attached to your bones, either directly or by thick, strong bundles of fibres called tendons. You can feel one of them, called the Achilles' tendon, in the back of your ankle. However, voluntary muscles do not just move limbs. Your tongue, eyes and jaw, for instance, cannot move without these muscles.

A microscope photograph of cells found in a muscle.

Your muscles are made up of thin, flexible fibres. Each fibre is connected to a nerve (see page 34), and the fibres in one muscle all work together. The nerve receives messages from your brain, telling the muscle when to contract (shorten) and by how much. When you do not need to use that muscle again, your brain tells all the nerves to relax (lengthen) the muscle fibres.

Your muscles are arranged in groups. The simplest kind of group is used to move hinge joints (see page 12). These have pairs of muscles that are called antagonistic, meaning that they work against each other. For example, if you feel the front of your upper arm, and move your hand towards your face, you will feel your biceps muscle. It bunches as the muscle contracts, pulling your forearm up. Now straighten your arm and feel the muscle relax and become flat. Do this again, but feel the triceps muscle at the back of your upper arm.

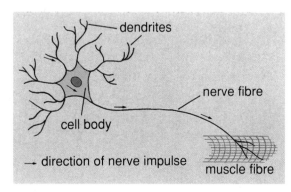

A diagram to show how a nerve cell is attached to a muscle fibre.

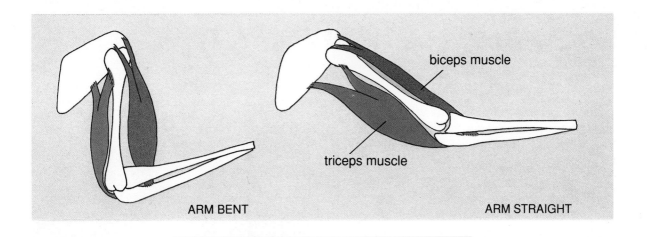

ARM BENT   ARM STRAIGHT

ACTIVITY

MAKING A MODEL ARM

YOU NEED

- **strong card**
- **2 strong rubber bands**
- **5 push-through paper fasteners**
- **2 sticky labels**

1 Cut out two shapes from strong card, as shown in the diagram.

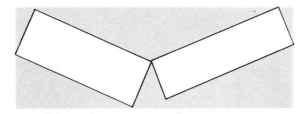

2 Join the shapes together with a paper fastener through the 'elbow'.

3 Connect the shapes together with the rubber bands. One band is the biceps and the other is the triceps. Label them with the sticky labels.

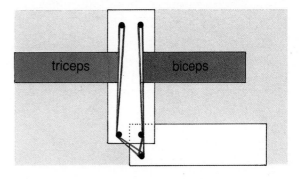

4 Bend the model arm. What happens to the biceps and the triceps?

5 Straighten the arm. What happens to the biceps and the triceps?

TEST YOURSELF

1. What are the two kinds of muscle in your body? Give an example of where they are used.
2. Why does your arm have two muscles to work the elbow joint?
3. Describe how a muscle works.

YOUR DIET

When you hear the word 'diet', you probably think of someone who is trying to lose weight. It really means the food that people eat in their everyday lives. People around the world have a great range of diets, but most of them contain the same basic food substances. The three main foods are carbohydrates, fats and proteins. We also need vitamins, minerals, water and roughage, or fibre. All these food substances are needed for energy, growth, repair and general good health.

When your muscles move, they are using energy. You need energy for everything that you do, even sleeping. You have to keep warm, and your heart must keep beating. Even your brain uses energy to make it work. You need it so that your body can perform all its tasks properly. It helps your body to grow and repair itself. Men need the most energy, because they are usually the largest. Women who are expecting

A chart showing the average amounts of energy required each day by people of different ages and sex.

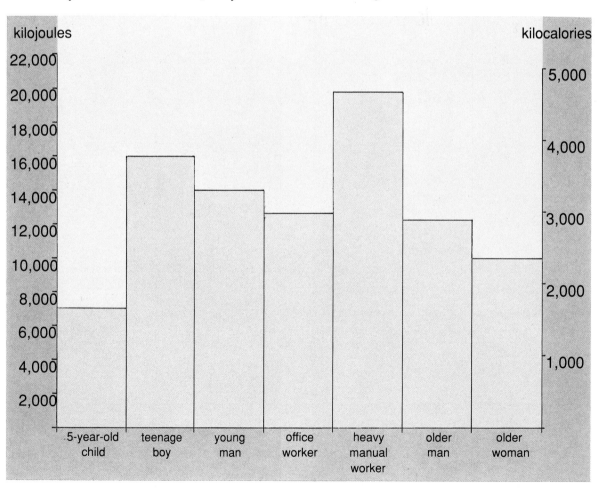

A healthy diet includes fresh fruit and vegetables, lean meat, fish, eggs, milk and yoghurt, wholemeal bread, wholewheat cereals and plenty of water to drink.

babies, or are breastfeeding them, need more energy than most other women. Children younger than twelve years old need less energy than those who are older. Children over twelve need roughly the same amount of energy as adults. Most of this energy comes from carbohydrates and fats.

The main type of carbohydrate is starch, which is found in rice, potatoes, bread, pasta, cassava, yams and a wide range of cereals. Sugars are also carbohydrates. Fats and oils are found in butter, milk, meat, cheese, fish and eggs. These are called animal fats. Vegetable fats and oils are found in seeds, such as sunflower seeds, and some vegetables, such as avocado. Many doctors and scientists think that people should eat only small amounts of fat, particularly that which comes from animals. They believe that high-fat diets cause overweight and heart disease.

Proteins are very important. They are needed to repair the body and help it to grow. They also provide some energy. They are found in fish, eggs, meat, cheese, milk, poultry and some vegetable products. People who do not eat meat are called vegetarians, and those that eat no animal products are called vegans. They get their protein from beans and pulses, such as lentils.

We need only tiny amounts of vitamins and minerals. Vitamins are found in fresh food. There are many types, and without some of them you would become very ill and eventually die. For example, your bones and teeth may not form properly if you do not have the correct vitamins when you are a child. The lack of certain other vitamins causes your eyes, skin and blood to become unhealthy. Minerals are also vital for life. They can be found in most normal, fresh-food diets. Some minerals you may have heard of are calcium (for bones and teeth) and iron (for the blood).

Water is also essential for life. About three-quarters of your body is water. It is found all over your body. Water in your diet comes from drinking and eating. You could be starved of food for several weeks, but, without water, you would die within a few days.

Most food substances contain material that cannot be digested (see page 23). This is called dietary fibre, bulk or roughage. It is very important to have plenty of this in your diet. Doctors and scientists think that it can help prevent cancer of the bowel, and it certainly stops you from becoming constipated. It is found mainly in fruit, vegetables and wholegrain products, such as brown rice and wholemeal bread.

Citrus fruits (such as oranges, lemons and grapefruit) are an excellent addition to any diet. They contain dietary fibre, a large amount of water and natural sugar for energy. They are best known for their high levels of vitamin C, which helps to fight colds and other diseases.

ACTIVITY

> **YOU NEED**
>
> - a range of different diet sheets
> - packets and tins from foods you have eaten

1. List all the foods that you have eaten in one day.
2. Use the diet sheets and the labels on food packages to find out which foods are carbohydrates and which are proteins or fats. Do any of these foods contain dietary fibre?
3. Half your food should be starch: the rest, protein with a little fat or oil. How balanced was your diet for that day?
4. Plan a menu for next week. List all the foods that you will eat. Have you repeated any meals?
5. Which day do you think contains the best balanced diet?
6. List your favourite foods. Are these foods good for you? Can they be included in your balanced diet?
7. Which of your friends eat different types of food from you? What is different about their food? Are their diets balanced?
8. Find out all you can about the diets of other nations.

Time	Monday Food eaten	Carbohydrate	Protein	Fat or oil	Dietary fibre
8.00	Toast				
	Butter				
	Marmite				
	Tea				
	Milk				
	Sugar				
11.00	2 Biscuits				
	Juice				
12.45	Fish				
	Chips				

TEST YOURSELF

1. What are the three main substances found in most diets?
2. What does your body need in order to carry out cell repair, replacement and growth?
3. Plan a menu for one day that you consider to be a good, balanced diet.

YOUR TEETH

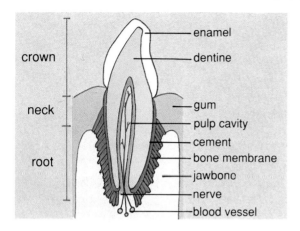

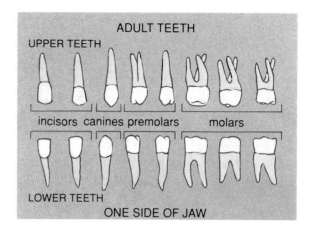

The food that you eat needs to be broken down into tiny pieces, so that you can swallow it easily and your stomach can digest it better. To do this, you have two rows of teeth, which bite, tear, cut and grind your food.

You have two sets of teeth in your life. The first set is called milk teeth, and there are twenty of them. You probably got your first tooth when you were about six months old. All twenty usually appear by the time a child is twenty months old. When you are about six years old, your adult teeth start pushing out the milk teeth, and extra ones grow until you have thirty-two.

Humans eat a wide range of foods, so they need different kinds of teeth. There are four main types of tooth. You have eight incisors at the front of your mouth, which are used to cut your food. Next to them, you have four canines ('dog' teeth). They are used for tearing. At the sides of your mouth, you have eight premolars and twelve molars, which have bumpy surfaces. Your food is ground into small pieces between the molars, and the premolars help with grinding and tearing.

Your teeth are alive. Each tooth is covered by a thin layer of enamel, which is the hardest substance that your body makes. Beneath this is a soft substance called dentine. This contains a pulp cavity, through which the blood vessels and nerves run. Each tooth has roots that fit into a socket in the jawbone. All the teeth are held in place by rubbery gums. The roots of your teeth are held in the sockets by ligaments that take the impact if you bite on something hard.

It takes only a few minutes every day to keep your teeth healthy and looking good. Most dental (tooth) disease starts from sticky plaque that covers the teeth. Plaque contains millions of bacteria that make the acid causing tooth decay. Sugar causes most decay, as it provides food for the bacteria to grow and spread. To prevent tooth decay and gum disease, you must brush and floss your teeth regularly. Most dentists say that you should brush your teeth at least twice a day, and visit a dentist every six months for a check-up. If you look after your teeth, they should last you for life.

ACTIVITY

CLEANING YOUR TEETH PROPERLY

> YOU NEED
>
> - **your own toothbrush**
> - **toothpaste**
> - **a cup**
> - **running water**
> - **disclosing tablets**
> - **a dentist's mirror**
> - **an ordinary mirror**

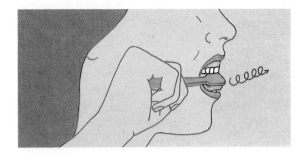

1. Suck a disclosing tablet. This will show up all the plaque on your teeth.
2. Look in the mirror. Use your dentist's mirror to examine your teeth. Notice all the areas of plaque that the tablet has stained.

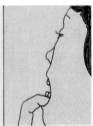

3. Spread some toothpaste on your brush.
4. Hold your brush so that it is at an angle to your gums. Brush near the top of your teeth, using little circular movements. Watch yourself in the mirror as you do it.
5. Do this for each tooth, in front and behind. Do not press hard.
6. Rinse your mouth. Look in the mirror. Have you managed to get rid of most of the stained plaque?
7. Brush over the top surfaces of your teeth, with a back-and-forth movement. Use more toothpaste if you want to.

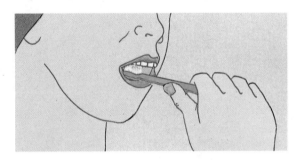

8. Rinse your mouth again. Look at your teeth. If you have not got rid of the stain, repeat steps 4 to 8, until you have cleaned your teeth properly.
9. How long did it take you to do this? Do you usually spend this much time brushing your teeth?

TEST YOURSELF

1. What are the four kinds of adult teeth? How many are there?
2. What causes teeth to decay?
3. What should you do to keep your teeth healthy?

YOUR DIGESTIVE SYSTEM

The food that you eat needs to be broken down into small chemicals that your body can use for growth, repair and energy. To do this, you have a digestive system, which is like a long tube running through your body. It has several parts, each with its own task.

Your digestive system contains or makes different chemicals in each part. These help the food to break down and dissolve in water. It has to do this so that it can pass through the gut walls to reach the places where it is needed. These chemicals are called enzymes, and each enzyme works on a particular type of food. Some enzymes work only in acid conditions, while others need alkali. Cells in the gut walls secrete (give out) acid or alkali: whichever is needed.

As you chew your food, glands in your mouth release saliva. This helps to soften the food, so that it can pass down your gullet, or oesophagus, into the stomach. Saliva also contains an enzyme that starts to break down starch. The stomach is the widest part of the digestive tract (tube). It is acid in the stomach, to help an enzyme break down proteins. The stomach also mashes the food into a soft, runny paste called chyme. This passes down into the top of the small intestine. Bile from the gall bladder in the liver breaks up oils and fats into tiny droplets. This makes them easier to digest. Another gland, called the pancreas, secretes an alkali to help enzymes to digest fats, proteins and starch. As the food passes further down the small intestine, more enzymes break the food down even more.

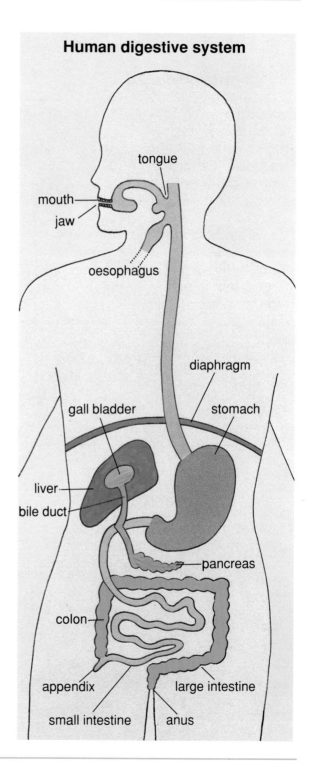

Human digestive system

Anything that is not digested and absorbed (soaked up) by the gut walls is waste. This is usually dietary fibre (see page 18). This builds up in the large intestine, is passed to the rectum by the squeezing of involuntary muscles (see page 14) and out through the anus.

ACTIVITY

ENZYMES AT WORK

YOU NEED

- **egg white**
- **fresh pineapple juice**
- **several biological washing powders**
- **a white cloth**
- **meat paste**
- **a covered glass jar**
- **egg yolk**
- **chocolate**

WARNING: in case you have sensitive skin, wear rubber gloves when using the washing powders.

1 Put a little egg white in one glass jar. Cover it with pineapple juice. Cover the jar to keep out dust.

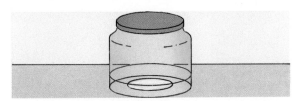

2 Shake the jar occasionally. Slowly the egg white will dissolve due to the action of enzymes in pineapple juice.

3 Take small pieces of white cloth and smear them with meat paste, egg yolk and chocolate.

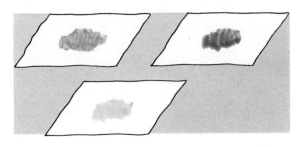

4 Read the instructions on the biological washing powder packets. Wash each stained cloth in warm water. Do the enzymes dissolve the food stains? Which stain is the most difficult to remove? Which biological washing powder works best?

TEST YOURSELF

1. Why do you have a digestive system?
2. What happens to food in your mouth?
3. What does your stomach do?
4. What do enzymes do?

FOOD ABSORPTION

You know that the chemicals from broken-down food have to be dissolved in water so that they can be absorbed through the gut walls. From here, they go into your bloodstream, to be taken to places where they are needed.

Carbohydrates, mainly starch and sugars, are broken down to glucose, which is a chemical that dissolves easily in water and can be absorbed quickly. Proteins are broken down to amino acids, which are also water-soluble. Fats and oils are broken down into tiny droplets, which 'hang' in water. This is called an emulsion.

The small intestine is the place where most of the chemicals are absorbed. It is very narrow and can be as much as 6 metres long. Its lining is not smooth, but is covered in millions of villi, which are tiny finger-like projections.

These villi greatly increase the surface area. This means that the surface area through which food chemicals are absorbed is far larger than it would be if it were smooth. The villi contain tiny blood vessels, which absorb the nutrients (food chemicals). Cells in the walls of the intestine take in the tiny fat and oil droplets.

All the absorbed food chemicals travel in the bloodstream to the liver. This is quite large and has many tasks. It checks the absorbed nutrients and if there is too much of a particular kind, the liver will store it. The liver also deals with poisonous or harmful substances. It will try to break them down to make them harmless. Nutrients that are not absorbed by the small intestine move on to the large intestine. Here, water, minerals and vitamins are absorbed.

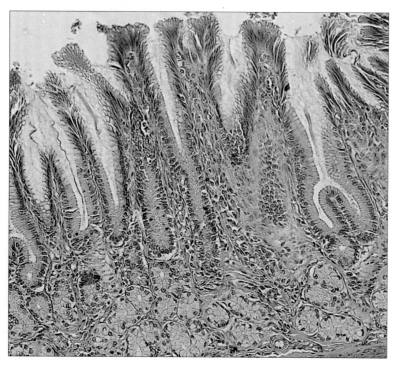

A microscope photograph of the lining of the intestine, showing the finger-like villi.

ACTIVITY
INCREASING SURFACE AREA

YOU NEED

- a marker pen
- Plasticine or modelling clay
- a ball of string
- a ruler
- a rolling-pin
- a kitchen knife

WARNING: take care when using the knife. It does not have to be a sharp one.

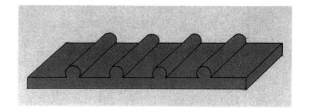

5 Measure across the flat piece of Plasticine, and record this length.

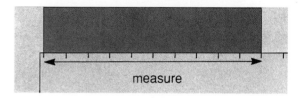

1 Use the rolling pin to roll out a fairly large ball of Plasticine.
2 Cut out two strips about 10 cm long and 2 cm wide. Make sure they are both the same length.

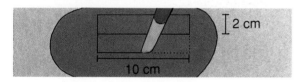

6 Cut off a length of string from your ball. It should be about 50 cm long.
7 Ask a friend to hold one end of the string at one end of the strip with the 'sausages'. Run the string carefully over the 'sausages', making sure that it touches the Plasticine. When you reach the other side of the strip, mark the string with the pen.

3 Use the rest of the Plasticine to make four 'sausage' shapes. They should be about as thick as your little finger, and about 2 cm long.
4 Stick the 'sausages' along one of your strips of Plasticine. This is a model of villi on the wall of the small intestine.

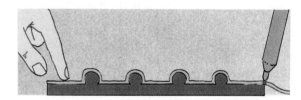

8 Take the string away, and measure it with your ruler. How does this length compare with the length across the flat strip? What have the 'sausages' done to the surface area of the Plasticine strip?

TEST YOURSELF

1. What are nutrients?
2. How do nutrients get into your bloodstream?
3. What does your liver do?

YOUR BLOOD

You know that the chemicals from your food need to reach your cells. Some of this food gives you energy by being 'burnt' in the cells with oxygen. That oxygen must be carried from your lungs to those cells (see page 30). You need to be protected from disease and wounds need to be healed. Waste has to be taken away from your cells. Your blood is able to do all these jobs.

Blood is made up of a watery yellow fluid called plasma, in which float millions of tiny cells. Its red colour comes from disc-shaped cells called red blood cells. There are over 5 million of these in 1 cubic millimetre of blood. They contain a chemical called haemoglobin. This picks up oxygen from your lungs and takes it to your cells. If your red blood cells are the wrong shape, or you do not have enough haemoglobin, you have anaemia. You look pale and become tired and weak very easily.

Your blood also contains several different types of white blood cell. Each cubic millimetre of blood contains 5,000-10,000 of these. They are all mainly used for defending your body

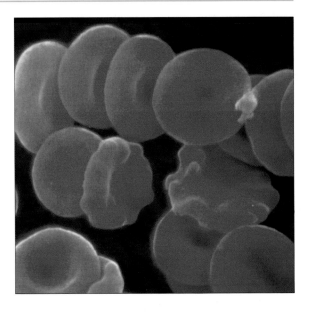

An electron microscope photograph (specially coloured) showing red blood cells.

against bacteria and other harmful materials. Some of the white blood cells do this by attacking the 'foreign' particles and engulfing (surrounding) them.

If you have a wound, cells called platelets help to plug the hole. A substance called fibrinogen makes a network of tiny fibres. It does this by turning into fibrin. A scab eventually forms, and drops off when the area of skin is mended. White blood cells rush to the wound, so that they can fight any bacteria or dangerous substances that come in through broken skin.

Your blood must be continually cleaned, otherwise you would become seriously ill and eventually die. Your kidneys perform this task. Every five

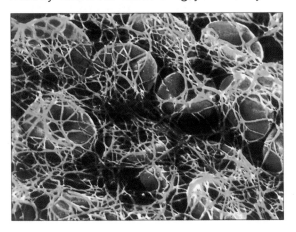

An electron microscope photograph showing a blood clot.

minutes, all your blood has passed through your kidneys and has been cleaned. These waste products that your kidney has filtered out of your blood collect, along with water, in your bladder. This is called urine, and is passed out of your body several times a day. An adult usually passes 1 to 1.5 litres of urine per day, but this amount may be as much as 3 litres.

ACTIVITY

HOW MUCH BLOOD DO YOU HAVE?

> YOU NEED
> - **bathroom scales, marked in kg**
> - **a calculator**
> - **a measuring cylinder**
> - **a washing-up bowl**

INFORMATION: An adult weighing about 70 kg has about 6 litres of blood. A small child weighing 12 kg has about 1 litre of blood. Therefore, roughly 12 kg of body weight is equivalent to 1 litre of blood.

1. Weigh yourself on the bathroom scales. Record your weight to the nearest kilogram.
2. To work out how many litres of blood you have, write down:
   your weight in kg ÷ 12

> Number of litres of blood =
> $$\frac{\text{MY WEIGHT IN KILOGRAMMES}}{12}$$

3. Use your calculator to find out the answer. Give that answer to the nearest 0.5 (half) litre. Write your answer under the calculation in part 2.

> $$\frac{\text{MY WEIGHT IN KILOGRAMMES}}{12}$$
>
> = ? LITRES

4. Use your measuring cylinder to see what the volume of your blood looks like. Work out how many times you will have to fill it with water. Do this, and pour the water each time into the washing-up bowl.

5. Compare your result with those of everyone in your class, by drawing a bar chart.

TEST YOURSELF

1. What do red blood cells do?
2. How do white blood cells fight 'foreign' particles, such as bacteria?
3. Describe what happens if you cut your knee.

YOUR CIRCULATORY SYSTEM

Your blood needs to be able to travel to all parts of your body, so that it can do its work properly. It is pumped round in tubes called vessels, which form a high-speed, one-way transport system around your body. This is called the circulation, or circulatory system. The pump that causes the blood to flow is the heart.

You have three types of blood vessel. The tubes that come out of your heart are called arteries. These divide into smaller and smaller branches, until they become tiny vessels called capillaries. These run very close to your cells. The chemicals from your food and the oxygen from your lungs pass through the thin walls of the capillaries. Waste materials from the cells, such as carbon dioxide produced by burning your food, pass back into the blood and are carried away. The blood returns to your heart in tubes called veins.

The entrance to the aorta. This artery takes blood from the heart to everywhere except the lungs.

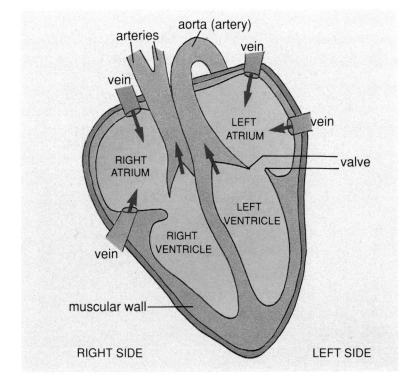

A diagram showing a cross-section of the heart.

Blood is very thick and sticky, so your heart has to pump it at quite a high pressure. Your arteries have very thick, elastic walls, so that they do not burst under this pressure. The pressure inside the veins is not as high as in arteries, so there are flaps called valves inside your veins. These work like gates, shutting if the blood tries to flow backwards.

Your heart is very muscular and strong, because it has to beat about once every second for the whole of your life. It is really two separate pumps, side by side. Each pump has two compartments: an atrium (plural: atria) and a ventricle. The ventricle is the main pump, so it has thick walls. The atrium receives blood from your veins. This blood has lost almost all its pressure, so the walls of the atrium are thin. There are valves between an atrium and a ventricle, so that blood does not flow backwards. There are also valves at the outlets of the ventricles.

There are two parts to your circulation. Blood from your right ventricle flows along an artery to the lungs, where it picks up oxygen. From there, it passes back in a vein to your left atrium. It flows into the left ventricle, where it is pumped out, still full of oxygen, to the rest of your body. It is carried by several different arteries. It comes back, having dropped its oxygen, to the right atrium, where it passes to the right ventricle. From here, the cycle starts again. The blood in your arteries going to your body contains oxygen, and the blood in your veins contains carbon dioxide waste, which is taken back to the lungs, to be breathed out, by the artery from your right ventricle.

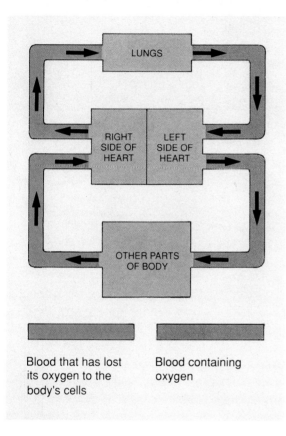

A simple diagram showing the pathway of the blood round the body. This pathway is called the circulation.

TEST YOURSELF

1. What are the three kinds of blood vessel?
2. What are your capillaries for?
3. Why do you have valves in some parts of your circulatory system?
4. Describe the pathway of blood round your body.

LUNGS AND BREATHING

You need oxygen, to burn food for energy, in the cells of your body. This process is called respiration, and carbon dioxide gas and water are the waste products. You get the oxygen from the air that you breathe in, and you breathe out the carbon dioxide and some of the water vapour.

When you breathe in, air goes into your mouth or nose and down a tube called the pharynx. Then it passes through your larynx (voice box) and into your windpipe, or trachea. This divides into two tubes called bronchi, which go into your two lungs. Inside your lungs, the bronchi divide into smaller and smaller branches called bronchioles. These end in tiny air sacs (bags) called alveoli. They greatly increase the surface area of the lungs. The alveoli are surrounded by tiny blood capillaries.

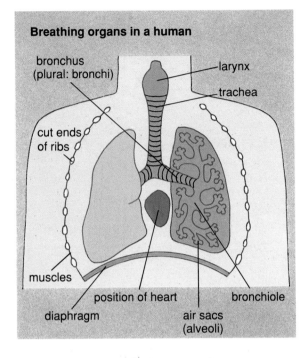

Breathing organs in a human

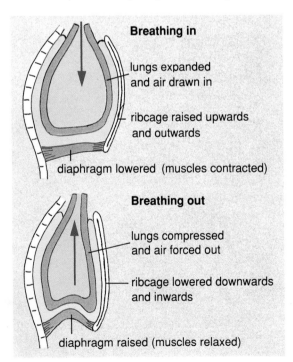

Your lung surfaces are wet, because the oxygen must be dissolved before it can pass to the rest of your body. When the alveoli fill with air, the blood capillaries pick up some oxygen and carry it away.

Below your lungs, there is a large sheet of muscle called the diaphragm. When you breathe in, your diaphragm contracts and flattens, and your ribs move up and out. This makes your chest cavity (space) larger, forcing your lungs to inflate (get larger) to fill the space. This makes air rush in. When you breathe out, the muscles between your ribs tighten. Your ribcage moves down and in. Your diaphragm relaxes and becomes dome-shaped. These two processes make your chest cavity smaller, forcing your lungs to collapse. This pushes the air out of your lungs, mouth and nose.

ACTIVITY

LUNG CAPACITY

> **YOU NEED**
>
> - a see-through plastic container, able to hold about 5 litres, with a lid
> - a large bowl
> - plastic tubing
> - a marker pen
> - antiseptic wipes

1. Put water in the bowl to a depth of about 5 cm.
2. Completely fill the plastic container with water.
3. Put on the lid. Turn the container upside down in the bowl.
4. Remove the lid underwater. This will keep the water in the container.
5. Insert the tubing under the rim of the container. Make sure that the top of the container stays underwater. Wipe the other end of the tube with an antiseptic wipe.

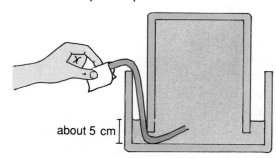
about 5 cm

6. Breathe in deeply. Hold your breath.
7. Put the end of the tube in your mouth. Breathe out, down the tube, until you have completely emptied your lungs.

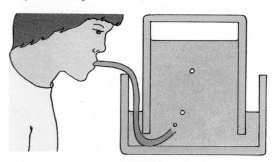

8. Pinch the tube as soon as you have finished, to prevent the water from rushing out.
9. Ask a friend to mark the water level on the side of the container. Write your initials beside the mark.

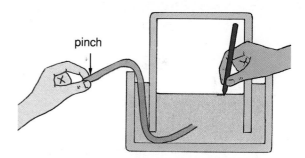
pinch

10. Compare your lung capacity with that of your friends. Who has the largest lung capacity? Does the smallest person have the smallest lung capacity?

> ### TEST YOURSELF
>
> 1. What is respiration? What are the waste products?
> 2. What happens in your chest when you breathe in?
> 3. What happens in your chest when you breathe out?

YOUR BRAIN

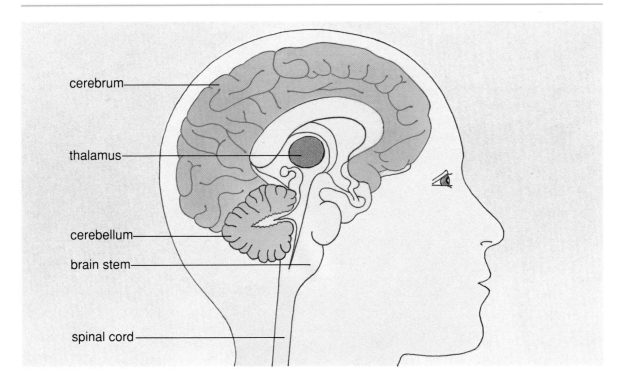

A diagram showing the main areas of the human brain: the cerebrum, the cerebellum and the brain stem.

Your brain controls your nervous system, which means that it also controls your whole body. For this reason, some people think of the brain as being like a complicated central computer. Most of the brain's control is involuntary (see page 34): for example, heartbeat, breathing, body temperature, muscle tension and other body systems. Your brain also controls your voluntary actions, such as running, walking, writing or reading. Besides these bodily actions, your brain helps you to learn, think, dream, imagine, remember and make judgements.

There are three parts to your brain: the cerebrum, cerebellum and brain stem. The cerebrum is the largest part and is found at the top of your brain. It is divided into halves called the cerebral hemispheres. Each half is in charge of the senses and movement in the opposite half of your body. For example, your left cerebral hemisphere controls the right side of your body. Both hemispheres control your breathing and swallowing. Your left hemisphere controls reading, writing, speech and any special skills that you have. However, if you are left-handed, your right hemisphere controls reading, writing and speech.

The front of your cerebrum is concerned with how you get on with other people, such as controlling your emotions and behaving properly in company. The back of your cerebrum tells you about light, shade, shape, colour and patterns. The lower part is for hearing, smell, anger, fear and sexual behaviour.

Your cerebellum is below the cerebrum and controls your posture, which is the correct positioning of the parts of your body, and helps movement. Your brain stem joins your cerebrum and cerebellum to your spinal cord. It controls the nerves of your head and neck. It also helps to control balance, breathing, blood circulation and wakefulness. All your nerve cells from the spinal cord pass through the brain stem. Complicated processes such as thought and memory are controlled by the whole of your brain, rather than by a particular part of it. The brain is so complicated that scientists and doctors may never understand exactly how it works.

This girl's brain plays a vital role in helping her to balance.

ACTIVITY

1 Choose a simple task, such as opening a door.
2 Perform that task. As you do it, think about every stage. What information are you receiving? What decisions do you have to make? What does your brain tell your body to do?
3 Write down everything that you have thought of. Make an 'instruction sheet', describing each stage of your task.
4 See if your friend can perform the task properly by following your instructions!

TEST YOURSELF

1. What are the three parts of your brain?
2. What does the left cerebral hemisphere control if you are right-handed?
3. What is the back of your brain for?
4. What does your brain stem do?

YOUR NERVOUS SYSTEM

Your nervous system is one of the most complicated parts of your body. Without it, you would not be able to survive, because it helps all the parts of your body to communicate with each other. It consists of your brain, spinal cord and the sensory and motor systems.

The sensory system is the part that picks up information, such as light from the eye. It is a vast network of nerve cells called neurones, which pick up stimuli (information) through nerve endings. Each stimulus is turned into an electrical pulse, which travels along the neurone. Usually, the electrical pulse is passed on to another neurone in the network. The neurones have fine branches at the end called dendrites, and these transmit (pass on) the information to the next neurone. Eventually, the pulse goes to your spinal cord, which is found inside your spine. This cord is connected to your brain, so that all information can be taken there to be interpreted.

Once your brain has found out what the electrical message means, it must tell your body how to react. This is when the motor system starts its work. Electrical pulses telling your body what to do are sent from the brain down the spinal cord. They pass into another huge network of motor neurones, which work in exactly the same way as the sensory neurones. The only difference is that the motor neurones usually end up in a completely different part of your body.

There are two types of action that your brain and nervous system cause. One is called voluntary, and means actions that you control, although you may not be aware of it. Some examples are walking, running and writing. Most of your actions are involuntary or automatic. This means that you cannot control them in any way. A good example of this is your 'reflexes'. You may have had these tested by the doctor; it is done to see if your nervous system is working properly. Cross one leg over the other in a sitting position, and get a friend to tap you gently with the side of the hand just below your

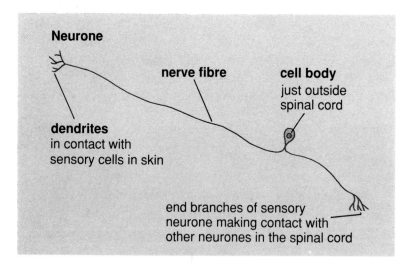

A diagram showing a sensory neurone. It takes messages from sensory cells to the spinal cord.

The woman is testing her 'knee-jerk' reflex. It shows whether that part of her nervous system is working properly.

kneecap. Your leg should jerk slightly. Another reflex action happens when saliva flows into your mouth if you are hungry and smell food.

Nerve cells are the only cells in the body that can never be repaired or replaced. That is why you react a little more slowly to events when you are older, and why some old people become forgetful.

ACTIVITY

INVOLUNTARY ACTIONS

YOU NEED

- a watch with a second hand
- an onion
- a knife
- a chopping board
- a well-lit room

1 Work with a friend. If you wear glasses, take them off.
2 Ask your friend to count how many times you blink in one minute. You can control this to some extent, but it is usually an involuntary action, so try not to think about it.
3 Cut an onion into small pieces on the chopping board. Get your friend to count how many times you blink in one minute. Has your blinking increased?

	number of blinks in 1 minute
without onion	
with onion	

Onion juice comes through the air as a vapour. It irritates your eyes, making them sting. Your brain tells your eyelids to blink faster, to bathe your eyes with tears that wash them. It also tells glands in the corners of your eyes to produce more tears, for the same reason. Many involuntary actions, like this one, are for safety.

TEST YOURSELF

1. What are the different parts of your nervous system?
2. What is a stimulus? Name two.
3. How does your motor system work?

YOUR SKIN

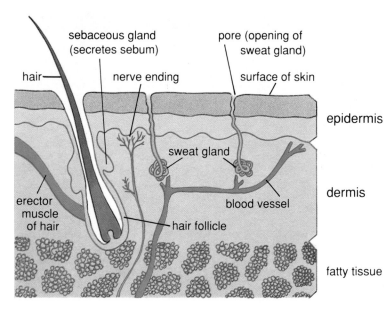

A diagram showing a cross-section of the skin.

This boy enjoys the softness of the hamster's fur using the nerves in his fingertips.

The whole of your body is covered by an organ called the skin. It is very important, because it protects the inside of your body from harmful substances, bacteria and ultraviolet rays from the Sun. It is waterproof, keeping your body fluids in and water out. It has two layers: the epidermis and the dermis. The epidermis is a thin, outer layer of cells, some of which are dead. The inner layer, or dermis, contains muscles, nerves, blood capillaries and glands.

Because you have to make so many different movements, your skin must be stretchy or it would split. It must be especially elastic over your joints, such as the elbow. Glands secrete (give out) an oily substance called sebum, which helps to keep the skin stretchy.

Your skin helps you to keep the temperature of your body correct. Most people's temperature is 37°C; if you get hotter or colder than this, even by a few degrees, you become ill. If you get too hot, your skin automatically cools you. Sweat glands secrete sweat. When this evaporates, it takes away the extra heat from your body, and you cool down.

When the outside temperature is cold, tiny blood capillaries constrict (close up), making your skin look blue. This cuts down the blood supply to your skin and stops heat from being lost. You also get 'goose pimples', which are caused when tiny hairs on your skin are raised by little muscles. The goose pimples are the little bunches of muscle. The hairs trap a layer of warm air next to your skin.

Your skin also contains nerves that tell you about your surroundings, through the sense of touch. You can feel heat and cold, as well as textures such as rough, smooth, hard and soft. Some parts of the skin have more nerve endings than others. For example, your fingertips have more nerve endings, so are more sensitive than the skin that covers your knees.

ACTIVITY

THE SENSITIVITY OF YOUR SKIN

> YOU NEED
> - a wide range of small objects with different textures
> - a blindfold

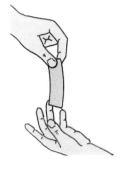

1 Work with a friend.
2 Make a chart showing the name of each different object and the areas of your skin that you are going to test. Some examples are fingertips, back of forearm, forehead, lips, knees and toes.

Area	Object 1	Object 2	Object 3
fingertips			
forehead			
lips			

3 Ask a friend to blindfold you.
4 Feel each object in turn with your fingertips. Your friend should mark a tick for each correct guess and a cross for each wrong answer.

5 Ask your friend to rub each object in turn against the back of your forearm. Do this in a different order from step 4.

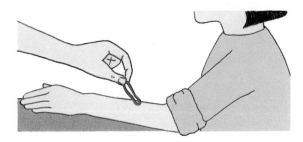

6 Repeat this for the other parts that you have chosen to test. Remember to get your friend to give you the objects in a different order each time.
7 Look at your results. Which is the most sensitive part of your skin that you tested? Which is the least sensitive? Can you think of reasons why?

TEST YOURSELF

1. Why must your skin be elastic? What helps to keep it elastic?
2. Explain how your skin is a barrier.
3. What happens to your skin when you are cold, and why?
4. What happens to your skin when you are hot, and why?

YOUR SENSES

Every moment of your life, even when sleeping, you need to find out about your surroundings, so that you can make decisions and avoid injury and death. To help you do this, you have five senses. You have already found out about your sense of touch (see page 37); the other senses are hearing, sight, taste and smell.

Your sense organs are all different, but they work in a similar way. They have sensitive cells that pick up information from the outside world. They pass this information to the brain, through nerves. Your brain then interprets the messages and tells your body what to do. For example, it might tell you to move your hand away from a hot surface. This all happens very quickly – in a few millionths of a second.

Your eyes are your organs of sight. Light is the stimulus that causes you to see. It passes through a transparent 'skin', called the cornea, at the front of your eye. Inside, it passes through a clear jelly to the lens, then through more clear jelly to the retina. This is where the light-sensitive cells are found. They send messages to your brain along the optic nerve. Your brain interprets these messages, telling you what you can see.

Your hearing organs are the ears. The stimulus is sound, which is pressure waves in the air. They cause a skin called the eardrum to vibrate, and this is passed on by a system of three tiny bones to your inner ear. Here, hairlike sensitive cells pick up the vibrations and turn them into electrical messages to be sent to your brain.

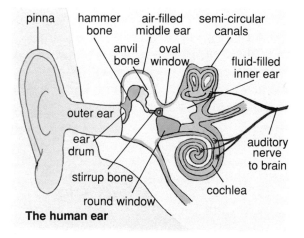
The human ear

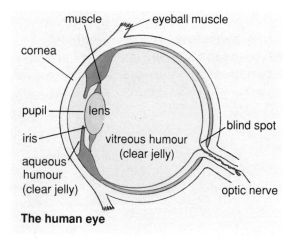
The human eye

A microscope showing the tastebuds on the tongue.

Your senses of taste and smell work in a similar way. Chemicals are the stimuli, and they dissolve in water either in your nose or on your tongue. Hollows in your nose have sensitive cells that detect the chemicals and send messages to your brain. On your tongue, you have taste buds, which are sensitive cells. Different areas detect salt, sour, sweet and bitter chemicals. They send messages to the brain. Your tongue and nose work together to help you to find out what you are eating or drinking.

ACTIVITY

SMELL AND TASTE

YOU NEED

- an apple
- a chopping board
- a pear
- an onion
- a blindfold
- a cup
- running water
- a kitchen knife

1 Work with a friend.
2 Very carefully cut up the onion, apple and pear into small pieces. Keep them separate.

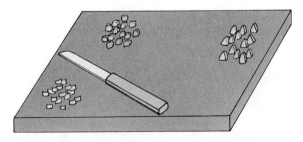

3 Ask your friend to blindfold you.
4 Get your friend to choose 2 different pieces of chopped food. Your friend must put one piece in your mouth and, at the same moment, hold the other piece under your nose. Smell and chew at the same time. What do you think you are eating?

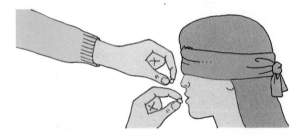

5 Rinse your mouth thoroughly with water. Repeat the experiment with 2 more pieces of chopped food.
6 Try all the combinations of smelling and tasting the chopped food.
7 Did you always guess correctly? Is one sense better than the other?
8 Repeat the activity with your friend blindfolded. Were your results the same as your friend's?

TEST YOURSELF

1. How many different senses have you got?
2. How are sense organs similar to each other?
3. Describe how one of your sense organs works.

HORMONES

You have already found out that your brain and nervous system control your body. However, many of your body's activities are also controlled by chemical 'messengers' called hormones. These are carried round the body in tiny quantities by the blood. Most hormones are made by special organs called endocrine glands. Some hormones affect every cell in your body, whereas others are used for only one particular task.

There are five main endocrine glands: pituitary, pancreas, adrenal, thyroid and parathyroid and sex glands (testes and ovaries – see page 42). The pituitary gland controls your growth, so it produces more hormone when you are a child than when you are grown up. The front of the pituitary gland is called the anterior lobe, and it controls your thyroid, adrenal and sex glands. The back (posterior lobe) controls the ways in which your blood vessels and parts of your kidneys work, and the breasts and uterus (see page 42) in women. Your pancreas controls the amount of sugar going to your cells.

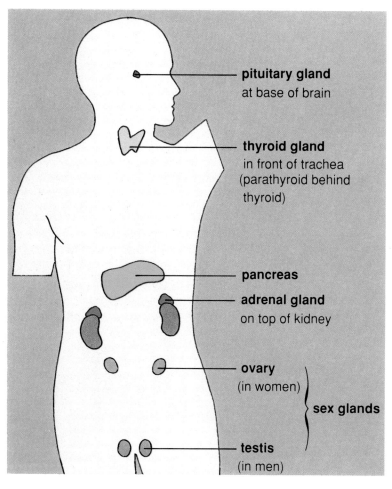

A diagram showing the main endocrine glands in the human body.

- **pituitary gland** at base of brain
- **thyroid gland** in front of trachea (parathyroid behind thyroid)
- **pancreas**
- **adrenal gland** on top of kidney
- **ovary** (in women) } sex glands
- **testis** (in men)

This girl is diabetic. The nurse is teaching her to inject herself with insulin. She will probably have to do this once a day for the rest of her life.

You may have heard of an illness called diabetes. A diabetic person's pancreas cannot produce a hormone called insulin. He or she may need injections of insulin, or a special diet, so that the correct amount of sugar goes to each cell. If the amount of sugar is too high or too low, the diabetic quickly becomes very ill and may even die.

Your adrenal glands have two parts: the cortex and the medulla. The cortex controls levels of salt in your body and helps to make your cells resistant to injury. It also helps to control the amount of protein (see page 16) that is built up in your body. The medulla makes a hormone called adrenaline, which prepares your body for quick action if something frightens you.

Your thyroid gland controls all your cells, helping them to use chemicals from food properly. There are four small parathyroid glands (attached to the thyroid), and they are the only glands that are not controlled by the pituitary. They seem to have only one task: controlling the amounts of calcium and phosphorus in your body.

TEST YOURSELF

1. What are hormones?
2. Which endocrine gland produces adrenaline? What is it for?
3. Which are the only glands that are not controlled by the pituitary?
4. What causes a person to be diabetic?

REPRODUCTION

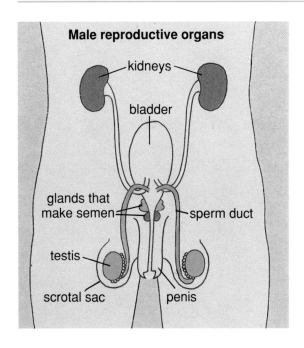

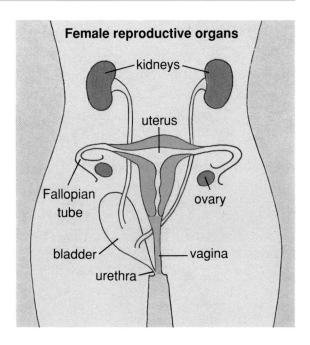

Humans so not live forever, so they must replace themselves. They do this by reproducing. When cells reproduce themselves (see page 8), they make an exact copy. If you look at your family, you will see that we do not make exact copies of ourselves. This is because children get a mixture of characteristics from both parents.

Men and women have sex cells. These carry information needed for the child to have all the correct parts of the body, to make sure that they work properly and to give the child its characteristics, such as eye and hair colour, height and even whether the child is good at certain things, such as mathematics. A baby girl is born with sex cells, which are stored in the ovaries. When she is about eleven years old (although this varies), she reaches puberty. This means that she begins to be able to reproduce.

Every month, one ovum (egg) is released from an ovary and travels down the Fallopian tube to the womb, or uterus. During that month, a lining of blood builds up in the uterus. Usually, the egg passes out of the girl's body, and so does the blood, about fourteen days later. This is called menstruation.

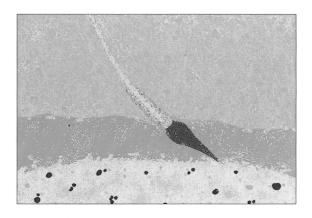

A sperm that has just passed through the outer layer of an egg during fertilization.

A baby boy is born with two testes. When he reaches puberty, they start to produce sex cells called sperm. These look rather like tiny tadpoles.

If a man and a woman want to have a baby, one ovum must join with one sperm. This happens during sexual intercourse. Here, the man's penis fills with blood, so that it can enter the woman's vagina. Millions of sperm are carried in a fluid called semen out of the end of his penis, and the sperm swim into the uterus and up the Fallopian tubes. In one of these tubes, the sperm meets the ovum, and one of them joins with it. This is called fertilization. The rest of the sperm die. The egg moves down the Fallopian tube and into the uterus. Here, it sticks to the blood lining, and develops into a baby.

The baby is attached to the lining of the womb by a special cord, which comes out of the abdomen. Your 'belly button' is the scar left by your cord. Food and oxygen pass through this tube to the baby, and waste products pass back to the mother. The baby floats in a liquid to stop it from being damaged by knocks.

After about forty weeks (nine months), the baby sends chemical messages that tell the mother's body that it is ready to be born. The uterus, which is very muscular, starts to contract (tighten), pushing the baby out through the mother's vagina. This is called labour. When the baby is born, its cord is cut and it starts to breathe.

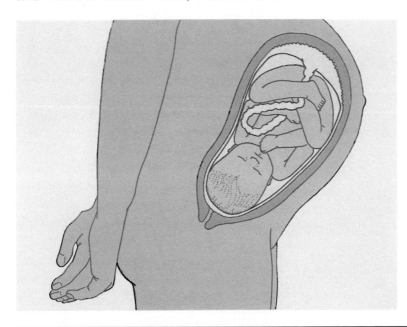

A diagram to show a baby inside its mother's uterus, shortly before it is ready to be born.

TEST YOURSELF

1. Where are a woman's sex cells found? Where are those of a man found?
2. Why are children not exact copies of their parents?
3. What is menstruation and why does it happen?
4. How does a baby inside its mother get its food and oxygen?

43

LEADING A HEALTHY LIFE

There are many things that you can do to make sure that you stay as healthy as possible. You have already learnt which foods you need to eat (see page 16). If you feel a bit 'run down', you can take extra vitamins and minerals. Your muscles and heart need to stay in good condition, so you should take exercise. Walking quickly for about twenty minutes, three times a week, is thought to keep you quite fit. Climb the stairs instead of using a lift. Swimming is very good exercise for the whole body.

It is very important to get plenty of sleep. Most adults need about six to eight hours of sleep every day, and children need more. People who keep missing out on the amount of sleep they need become tired, irritable and cannot think properly. This means that they cannot do their work, and do not enjoy life. It is also a good idea to avoid alcohol, smoking and dangerous drugs.

There are bacteria all around you. Most of them are harmless, but some will give you diseases. There are other ways to become ill: for example, some cells in your body may start to work in the wrong way, giving you cancer. Tiny organisms (living things) called viruses may invade your body and give you diseases like 'flu. Some of them are very dangerous indeed. You have probably had injections to stop you getting some diseases, and others can be cured by your doctor. If you feel ill, you should see a doctor.

Some people go to the doctor for regular check-ups, to make sure that their bodies are in good health. The doctor will weigh you, check your blood pressure and pulse, listen to your heart and lungs and sometimes do other tests, such as blood tests. The doctor may pick up warning signs of ill health, and will tell you what to do about it.

It is very important to your health to get enough sleep.

You should also visit your dentist regularly, to prevent or deal with tooth and gum disease.

Finally, remember to keep your body clean. Prepare, eat and store your food in clean places. Make sure that the food you buy is fresh, and always look for the 'sell by' date on packets and tins. Keep your home clean and get plenty of fresh air. Remember: you only have one body, and you will have a better life if you keep it healthy.

This little girl is unwell. Her father is giving her some medicine prescribed by the doctor. It is important to go to the doctor if you are ill.

ACTIVITY

A HEALTH SURVEY

> YOU NEED
>
> • **several friends**

1 Make up a survey about health. You could ask people how much sleep they get, how often they are ill, how often they go to the dentist, how often they brush their teeth and what exercise they take. These are just a few examples.
2 Give your survey to several friends.
3 Use the information on page 44 to see if your friends lead a healthy life.

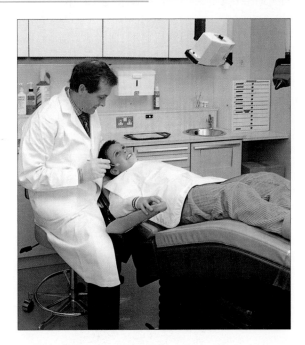

It is important to visit the dentist regularly.

TEST YOURSELF

1. How can you keep your heart and muscles in good condition?
2. What sort of tests will your doctor do at a general check-up?
3. What happens to people who do not get enough sleep?

Glossary

Abdomen The part of the body (trunk) between the diaphragm and the pelvis.
Acid A solution that is below pH7.
Alkali A solution that is above pH7.
Biological washing powder A washing powder that contains enzymes.
Bladder An organ that is like a bag. It usually means the organ that stores urine in the body.
Bowel The lower part of the digestive system.
Breastfeeding Giving milk to a small baby from the breasts, rather than from a bottle.
Calcium An essential element in bones and teeth.
Cancer A dangerous growth of cells in the body which in some cases can eventually cause death.
Cavity A hole or space.
Constipation This is caused when the bowels are not working properly, and waste material is stored in the body for too long.
Dental floss A special thread used for removing pieces of food from between the teeth.
Diaphragm A large muscle that separates the chest from the abdomen.
Digestion The dissolving of food in the stomach.
Fibre A fine thread.
Fibrin Fine threads formed in the blood when it is exposed to air. It is responsible for blood clotting.
Immunization A treatment to prevent disease, usually given by injection.

Kidneys Two small organs, at the back of the body, that are used to filter and clean the blood. They make urine.
Ligaments Tough, fibrous tissue that holds bones together.
Lens (of the eye) A clear material that focuses light rays on to the retina.
Optic nerve The bundle of nerves that sends electrical messages from the eye to the brain to be interpreted.
Organ A part of the body that carries out a particular task.
Oxygen A gas that makes up about 20 per cent of the air. It is essential for respiration in animals and humans.
Protein A complex substance containing nitrogen, carbon, hydrogen, oxygen and usually sulphur and phosphorus, that is essential for growth and repair in the body.
Pulse A beat that can be felt under the skin. It corresponds to the heartbeat.
Reflex An involuntary or automatic nervous reaction.
Tendon A fibrous material holding muscles to bone.
Tissue A collection of cells that are similar. Several different types of tissues may make up an organ.
Trunk The body of an animal or human, not including the head or limbs.
Urine A liquid that contains waste products. It is made by the kidneys.
Wholegrain products Foods that are made from the complete grain. No parts have been removed. They are usually high in dietary fibre.

Books to read

The Human Body Linda Gamlin (Franklin Watts, 1988)
The Human Machine Brenda Walpole (Wayland, 1990)
Pocket Book of the Human Body Brenda Walpole (Kingfisher Books, 1987)
The Young Scientist Book of the Human Body S. Meredith, A. Goldman and T. Lissaner (Usborne, 1989)
Healthy Living series (Wayland)
The Body in Action series (Franklin Watts)
You and Your Body series (Wayland)

Picture acknowledgements

The author and publishers would like to thank the following for allowing illustrations to be reproduced in this book: Ron Boardman 8, 12, 14, 24, 28, 38; St. Mary's Hospital Medical School 11; Science Photo Library 26, 41, 42; Topham *cover* (left); Wayland Picture Library *frontispiece*, 17 (Trevor Hill), 18 (Peter Stiles), 34, 36, 45 (top/Paul Seheult) 45 (bottom/Trevor Hill); ZEFA *cover* (top right), 44. All artwork is by Jenny Hughes.

Index

Air 6, 30, 45
Arteries 28–29

Bacteria 20, 26, 36, 44
Biceps 14, 15
Bladder 27
Blood 26–27, 28–29
Blood vessels 8, 20, 24, 28
Bones 10–11, 12–13, 14, 18
Brain 6, 10, 14, 32–33, 34, 38, 39, 40

Capillaries 28, 30, 36
Carbohydrates 16–17, 24
Carbon dioxide 28, 29, 30
Cartilage 13
Cells 8–9, 24, 26, 28, 38, 39, 40, 41, 42, 44

Dentist 20, 45
Diet 16–17, 18–19
Dietary fibre 16, 18, 23
Digestive system 14, 22–23

Ears 38
Energy 8, 16–17, 22, 26
Enzymes 22–23
Eyes 14, 38

Fallopian tubes 43
Fats 16–17, 19, 22, 24
Food 6, 8, 14, 16–17, 18–19, 20, 22, 24, 26, 28, 41, 43, 45

Gall bladder 22
Glands 36, 40–41
Glucose 24
Gut 6, 22, 23, 24

Hearing 38
Heart 6, 10, 14, 16, 28–29
Heart disease 17

Insulin 41

Joints 12–13, 14, 36

Kidneys 6, 26, 27

Ligaments 13, 20
Limbs 6, 11, 14
Liver 6, 22, 24
Lungs 6, 10, 26, 28, 29, 30–31, 44

Minerals 16, 18, 24, 44
Muscles 8, 10, 14–15, 16, 36, 44

Nerves 14, 36, 37
Nervous system 32, 34–35, 40
Neurones 34

Oils 17, 19, 22, 24
Ovaries 42
Oxygen 26, 28, 29, 30, 43

Pancreas 22
Plaque 20, 21
Platelets 26

Proteins 8, 16–17, 19, 22, 24

Red blood cells 8, 25
Reflexes 34, 35
Ribcage 10, 30
Sex cells 8, 42–43
Sight 6, 38–39
Skeleton 10–11, 12
Skin 8, 36–37
Skull 10, 12
Smell 6, 38–39
Sperm 43
Spinal cord 10, 33, 34
Spine 10, 11, 12
Starch 17, 19, 22, 24
Stomach 20, 22
Sugars 17, 20, 24
Sweat 36

Taste 6, 38–39
Teeth 18, 20–21, 45
Tendons 14
Testes 43
Tongue 14
Tooth decay 20, 45
Touch 6, 36
Triceps 14, 15

Urine 27
Uterus 40, 43

Vagina 43
Valves 29
Veins 28–29
Vitamins 16, 18, 24, 44

White blood cells 12, 26
Windpipe 30

48